The Surprise Party

Hodder Cambridge Primary Maths Foundation Stage

Ann and Paul Broadbent

This book belongs to

....

HODDER EDUCATION
AN HACHETTE UK COMPANY

Today is Max's birthday.
It will be full of surprises.

What is this present?

Look at the patterns on the presents.
Can you see a stripy pattern?
Which present has squares and circles in a pattern?

2

But another surprise is waiting!

Talk about the solid shapes.

What solid shapes can you name? Which presents are cubes? Which is a sphere? Can you find the lids for the open presents?

Read the numbers to 20 in order.

Which number comes after 11? Count on 3 from 14; which number do you land on? Which number comes before 15?

Max decides to put his shoes in box 12.

Practise counting in 2s to 20.

Count the shoes in 2s. How many shoes are on the shelves?
How many shoes are on the floor?

Count the pins to make sets of 10.
How many pins are knocked over in lane 13?
How many more pins will make 10?

Find the difference in the number of balls on the shelves.
How many more blue balls are there than red balls?

Share the food items between two.

Can you see half a pizza? How many tomatoes will Max get if they are shared between 2? Share the cucumber sticks between 2.

It has been a great party.
But another surprise is waiting.

Talk about measures.

Which glass has the most juice? Can you put the bottles in order from most full to least full?

Corky the clown is here! He makes everyone laugh at his silly jokes.

Wow! What a fantastic surprise!

Talk about the shapes on the clown.
Name the flat shapes you can see. Can you see any triangles?
What are the shapes of his buttons? What shape is on his bow tie?

They all clap and cheer at Corky's clever tricks.

Compare the length of different objects.
Which balloon is the longest?
Is it longer than the clown's shoes?

But the biggest surprise of all is …

Talk about shape and symmetry.
Is the bow tie symmetrical?
Can you see any other symmetrical shapes?

... the clown is DAD!

Talk about the time.

Can you see what time it is? Go back and find all the clocks in the story. Can you say the time on each page?

Write the missing numbers on this grid.

1	2	3	4	5
6	7	8	9	10
11	12	13	14	15
16	17	18	19	20

 Count the shoes in each row.
Draw more shoes to make a total of 10.

Add these balls together and write the totals.

5 + ☐ = ☐

6 + ☐ = ☐

 Colour 3 balloons green. Colour the rest yellow.

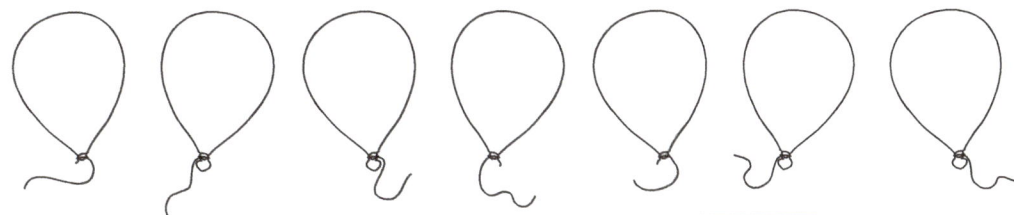

How many balloons are yellow? ☐

How many balloons are there altogether? ☐

Colour the cubes red.
Colour the spheres blue.

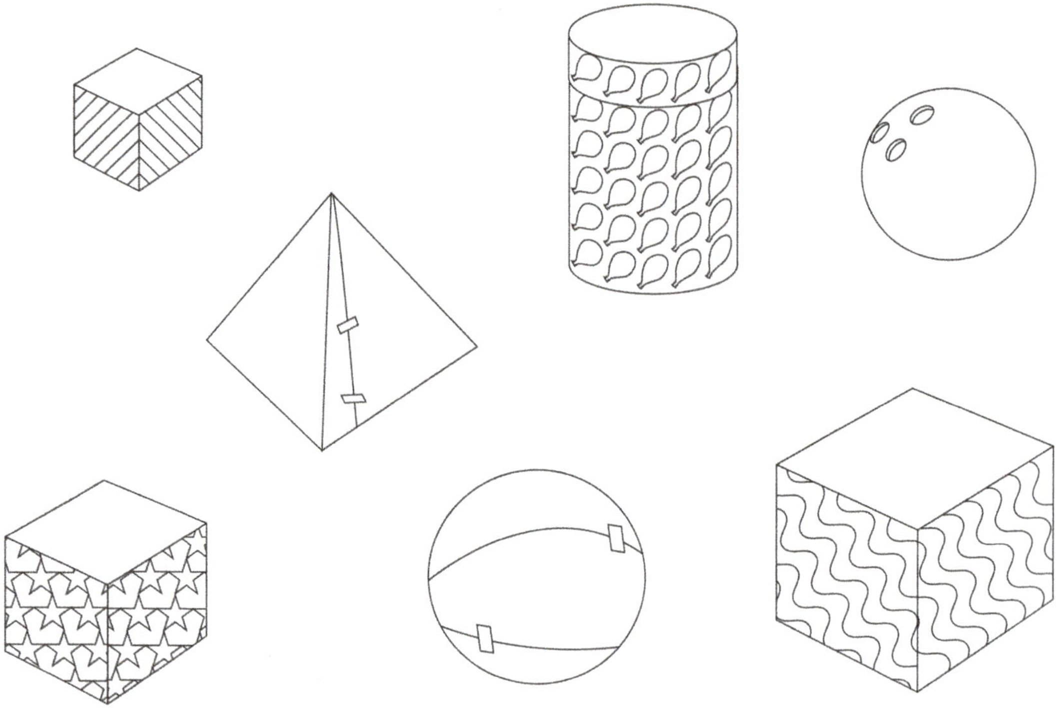

 Draw 3 more shapes for each pattern.
Colour each set of shapes to make a pattern.

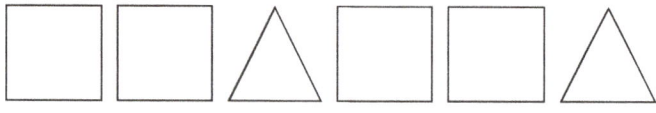